Kandan Balaji

Management of Coconut Eriophyid Mite

AF153280

Kandan Balaji

Management of Coconut Eriophyid Mite

LAP LAMBERT Academic Publishing

Impressum / Imprint

Bibliografische Information der Deutschen Nationalbibliothek: Die Deutsche Nationalbibliothek verzeichnet diese Publikation in der Deutschen Nationalbibliografie; detaillierte bibliografische Daten sind im Internet über http://dnb.d-nb.de abrufbar.

Alle in diesem Buch genannten Marken und Produktnamen unterliegen warenzeichen-, marken- oder patentrechtlichem Schutz bzw. sind Warenzeichen oder eingetragene Warenzeichen der jeweiligen Inhaber. Die Wiedergabe von Marken, Produktnamen, Gebrauchsnamen, Handelsnamen, Warenbezeichnungen u.s.w. in diesem Werk berechtigt auch ohne besondere Kennzeichnung nicht zu der Annahme, dass solche Namen im Sinne der Warenzeichen- und Markenschutzgesetzgebung als frei zu betrachten wären und daher von jedermann benutzt werden dürften.

Bibliographic information published by the Deutsche Nationalbibliothek: The Deutsche Nationalbibliothek lists this publication in the Deutsche Nationalbibliografie; detailed bibliographic data are available in the Internet at http://dnb.d-nb.de.

Any brand names and product names mentioned in this book are subject to trademark, brand or patent protection and are trademarks or registered trademarks of their respective holders. The use of brand names, product names, common names, trade names, product descriptions etc. even without a particular marking in this work is in no way to be construed to mean that such names may be regarded as unrestricted in respect of trademark and brand protection legislation and could thus be used by anyone.

Coverbild / Cover image: www.ingimage.com

Verlag / Publisher:
LAP LAMBERT Academic Publishing
ist ein Imprint der / is a trademark of
OmniScriptum GmbH & Co. KG
Heinrich-Böcking-Str. 6-8, 66121 Saarbrücken, Deutschland / Germany
Email: info@lap-publishing.com

Herstellung: siehe letzte Seite /
Printed at: see last page
ISBN: 978-3-659-78501-6

TABLE OF CONTENTS

MANAGEMENT OF COCONUT ERIOPHYID MITE *Aceria guerreronis* Keifer (Acari: Eriophyida)

INTRODUCTION

The coconut palm, *Cocos nucifere* L. is one of the plantation crops in the world. In Sansktrit, it is called 'Kalpavriksha' 'the tree of heaven or "the paradise tree 'which provide all the necessities of life (Daniel Sundarajan and Thulasidas, 1993; Nampoorthi, 1999). Coconut is extensively cultivated in 80 countries of the world with a total production of 54,129 million nuts from an area of about 120 million hectares (Nair and Rajesh, 2001). India is in the forefront among the coconut growing countries in the world. In India, this plantation crop is grown in 1.78 million hectares producing 12.252 million nuts with productivity of 6892 nuts per hectares per year during 1990-2000 (Singh, 2000). This crop contributes Rs.7000 crores to the gross domestic product of the country and earns valuable exchange to the extent of 313 crores by way of export of coir and coir products. Six per cent of vegetable oil consumption is met out by this crop. In India, Tamil Nadu ranks third in area with 266.50 thousand hectares, second in production with 3096.70 million nuts and fifth in productivity with 11620 nuts per hectare during the year 1998-1999 (Nair and Rajesh, 2001).

Among the various non insect pests that have been reported on coconut palm, eriophyid mite, *Aceria* [Erophyses] *guerreronis* Keifer (Acari: Eriophyida) is a serious one in Southern States in India. They generally suck the sap from the meristematic tissue of the nuts resulting in the loss of production of nuts (Kannaiyan *et al*, 2000). An eriophyid phytophagus mite, *Aceria guerreronis* Keifer was first described in 1965 from coconuts of Guerrero State, Mexico (Keifer, 1965). In the Indian sub continent, it was first reported from Srivilliputhur area of Tamil Nadu 1984 (Sathiamma *et al*, 1998).The estimated average loss in copra yield due to mite infestation was found to be 10-15 per cent in Tamil Nadu as compared to 10 per cent in Mexico and 11-18 per cent in St. Lucia (Moore and Howard, 1996).

The coconut mite was found in tropical and subtropical climates, but populations could survive both short period of frost and period of temperature just above 0°C (Zuluaga and Sanchez1971).The presence of mites in the nut was evident round the year with a slight reduction during the rainy season (Subaharan *et al*, 2001).

The control measures currently employed, including the foliar spraying and root feeding of chemical pesticides have proved to be partial success. Using synthetic pesticides has been beset with many problems causing environmental pollution and health hazards. Alternative control measures involving the use of eco friendly bio control molecules like botanicals and neem based pesticides within the ambience of integrated pest management principles have become imperative (Ramarethinam *et al*, 2000b). Above said information's laid the basis for formulating the following objectives of the present studies.

- To survey the incidence of *A.guerreronis* in the coastal belts of Tamil Nadu
- To study the most preferred stage of nut *A.guerreronis*.
- To study the population dynamics of *A.guerreronis*.
- To determine the efficacy of certain plant products against *A.guerreronis*.

REVIEW OF LITERATURE

The literature pertaining to the research on "Management of coconut eriophyid mite, *A.guerreronis* (Keifer) (Acari: Eriophyidae) "is reviewed under the following headings.

1.1 Prevalence and the spread of the pest

1.1.1 International prevalence

Damage symptoms of the mites were described from Colombia by Martyn (1949).The pest was reported from Colombia in 1971 (Zuluaga and Sachez, 1971). First report of damage caused to coconut by *A.guerreronis* in a Caribbean islands was reported by Estrada *et al* (1975). However *A.guerreronis* was first discovered in 1967 on coconut in Benin, as reported by Mariau (1977). It was also reported that the mites rapidly spread in Togo, Cameroon, Sao Tome, Nigeria and Ivory Coast, other reports of pest incidence were reported from Dominica (Moore and Alexander,1985),Veracruz, Mexico (Olevera and Fonseca,1986),Costa Rica (Schliesske,1988) and Mozambique (Uaciquete *et al*, 1998). The occurrence of the coconut mite also noticed in Hong Kong, Malaysia, Philippines, Singapore and Taiwan (Kannaiyan *et al*, 2000).

1.1.2 National prevalence

Sathiamma (1981) studied the mite fauna associated with coconut palm in Kerala and reported that surveillance is being maintained in India. So as to check *A. guerreronis* at the point of entry itself. Even though there was an alarm on detection of these mites in North Kerala and though the area was combed for its occurrence, this mite could not be traced out them. First report of the outbreak of nut infesting eriophyid mite was Ernakulam district of Central Kerala. Subsequently the devastating effects of these mites were reported from Bangalore in Karnataka, Udumalpet in Tamil Nadu (Sathiamma, *et al.*, 1998). Three islands of Lakshadweep namely Minicoy, Kalpeni and Kavaratti (Haq, 1999b) and Chittoor in Andra Pradesh (Shivarama Reddy and Naik, 2000). Now the pest has established in Kerala, Tamil Nadu and parts of Karnataka and Andra Pradesh.

1.1.3 State Prevalence

The occurrence of this mite was observed in Srivilliputhur in Tamil Nadu during 1984 and also from Pollachi and Udumalpet taluks of Coimbatore district in Tamil Nadu during August 1998 (Sathiamma *et al*, 1998). This mite has been reported to cause damage in the districts of Dindugal, Erode, Vellore, Thanjavur, Thiruvarur, Nagapattinam, Viruthunagar, Tirunelvelli (Kannaiyan *et al.*,2000).

1.2   Studies on bunch preference by *A.guerreronis*

Preference for colonization by the mites reported to vary with the age of nuts. Moore and Alexander (1987) found that the mites were not seen in unfertilized flowers but were present within a few weeks of fertilization. Infestation was very low in the first month after fertilization, but builds up rapidly to a peak on buttons of third bunch from the top and then dropped. The population tended to be low from ninth bunch.It appeared that bunches of one to four weeks old were the most susceptible to colonization.

Sathiamma *et al.* (1998) observed that nuts up to nine months of age harbored the mite, but fully mature nuts never contained any stage of mites. Colonies of mites comprising of eggs, first nymph, second nymph males and females were detected on nuts of six weeks age and the mite population showed a rapid increase during this period. This trend continued during progressive development of nuts up to the age of twelve weeks, followed by steep reduction during subsequent weeks and the population receded to a minimum level on nuts of the age of 22 to 24 weeks. Such difference in the duration of infestation may be due to varietal or age difference of palms or other ecological factors (Haq, 1999a).

Ramaraju *et al.* (2000) reported that population was very high in two to six month old buttons. He observed that two to forty mites along with large number of eggs in area of four square millimeter on infesting nuts. There were no mites in male flowers and also in female flowers before pollination. In some nuts, which had very soft husk with a lot of moisture, heavy attack was seen even month old tender nuts .otherwise, maximum population of mites were seen on third, fourth and fifth bunches from the fully opened flower bunches (Ranjith *et al.*, 2000). The mite population was found to be very high in 2-5 months old buttons (Kannaiyan *et al.*, 2000). Occurrence of large number of colonies of mites on 30-45 days old buttons was reported by Shivarama Reddy and Naik (2000).

1.3 Studies on population dynamics of *A.guerreronis*

Studies conducted on population dynamics from September 1998 to November 1999 at Agricultural Research Station, Aliyar Nagar, Tamil Nadu revealed that the population of eriophyid mite was maximum during May ($86/4mm^2$) followed by April ($73/4mm^2$) and March ($70/4mm^2$). The population was present even during rainy months (October-December, 1998) (Kannaiyan *et al.*, 2000).

Haq (1996b) reported that the number of mites per square centimeters area of the meristematic zone during the peak period of infestation reached more than 1200 under Kerala conditions. Infested nuts of 3 to 4 months of age from Sri Lanka and Lakshadweep showed an average of 600-8000 and 200-300 mites per Square centimeter of the meristematic zone respectively.

1.4 Management practices against *A.guerreronis*

1.4.1 Chemical control

Mariau and Techibozo (1973) reported that the best results were obtained with sprays containing 0.0125% quinomethionate, 0.04%, monocrotophos or 0.03%, tricyclohexatin hydroxide, which reduced the percentage of nuts infested from 82.0 to 98.8. They also reported that monocrotophos and quinomethionate were the best of 23 compounds tested for the control of *A.guerreronis* on coconut palms. Effective control was achieved by treatment with 0.0125% and 0.04% monocrotophos respectively, every three weeks. Mariau (1977) reported that spraying of

4

monocrotophos 0.04% at an interval of two months resulted in the reduction in nuts loss by about 90 per cent.

Hernandez (1977) best results were obtained with a spray containing 1-2 ml dicrotophos per litre applied at intervals of 15-20 days to the inflorescence and to fruit less than three month old. In1973-74, the effective control obtained with sprays containing 2ml monocrotophos or quinomethioanate per litre applied every 20 days and also reported that the best results were obtained with monocrotophos, dicrotophos, quinomethionate and cyclohexatin (Plictron) all at 1.5 ml or 1.5 g per litre equally good results were obtained with intervals of 20 or 30 days between treatments.

Julia and Mariau (1979) observed in pesticide trials promising results were obtained with quinomethioanate, monocrotophos and cyclohexatin. Julia *et al* (1979) observed that of the several pesticides tested in sprays against *E.gurreronis* (*A.gurreronis*) quinomethionate, monocrotophos, and cyclohexatin applied at 40-80 kg per ha every 3-9 weeks were found promising. Monocrotophos was used to control several eriophyoid species (Childers, 1996), including *A. guerreronis* on coconut (Mariau, 1977); Fernando *et al.*, 2000; Ramaraju *et al.*, 2000. Nair *et al* .(1999) observed that spraying of monocrotophos 4ml/litre of water recorded 78.20 per cent reduction in infestation. Muthiah and Baskaran (1999) reported that spraying methyl dematon 4ml/litre at 10 days interval effectively reduced the damage by mites (24.86%) followed by monocrotophos 1.5 ml/litre at 10 days interval (25.12%). Ramaraju *et al.*, (1999) suggested that spraying methyl -o dematon or triazophos or monocrotophos was found to be effective against the coconut mite *A. guerreronis*.

Four field experiments were conducted in Tamil Nadu Agricultural University, Coimbatore, two in the Agricultural Research Station, Aliyarnagar, and two in farmers holding. one in each at Avulur (Erode district) and Thathur (Coimbatore district) to evaluate the bio efficacy of insecticides and acaricides against the mite during October –December 1998, two rounds of spraying were gin and observations were made on the mite population .Among the treatments triazophos 5 ml/litre, monocrotophos 1.5 ml/litre and methyl dematon 4ml/litre recorded 70.30, 57.98 and 72.50 per cent mortality respectively (Ramaraju *et al.*, 2000).

Shivarama Reddy and Naik (2000) recommended that spraying chemicals *viz.*, dicofol 6ml/litre or 0.03% dimethoate for controlling the pest, Subaharan *et al.*, (2001) observed that spraying of monocrotophos 4 ml/litre of water recorded 78.15 per cent reduction in infestation.

Muthiah and Baskaran (2000) observed that six insecticides at higher concentration 5 ml/litre were evaluated for their efficacy against coconut mite along with fish oil rosin soap and monocrotophos root feeding. Among the chemicals tested, monocrotophos 36 SL and carbosulfan 25EC at 5 ml/litre were found to be highly with minimum mite population and maximum undamaged buttons of 100,100 and 73.61 per cent respectively four months after the first spray. Muthiah and Baskaran (2000) reported that the experiment conducted at Coconut Research Station, Veppankulam with the spraying of either methyl demeton 4ml/litre or monocrotophos 4ml/litre at 10 days interval significantly reduced the percentage of nut damage to 25 per cent compared to 53 per cent in untreated control.

Dey and Somchoudhury (2001) reported that efficacy of different pesticides against the eriophyid mite. When the pesticides were applied on the crown region, the population starts decreasing significantly 2.5% over control in all pesticidal treatments. At 8 days after treatment, significantly highest per cent reduction of mite population was recorded in case of fenazaquin 250ml/100 litre (92.77%), monocrotophos 250ml/100litre (70.51), dicofol338/100litre (73.47%). They were also reported that spraying of monocrotophos 2.5 ml/litre recorded 70.51 reduction of mite population at 8 days after treatment.

1.4.2 Botanicals on coconut mite

Ramarethinam and Marimuthu (2001) suggested that neem oil spray at the rate of 25 and 30 ml/litre of water to control *A.guerreronis*. Ramaraju *et al* (1999) observed that in the crown spraying experiment conducted at Avalur, TNAU neem oil 60 EC was superior to all other treatments recording the highest per cent mortality of 58.57 per cent at 7 days after treatment.

Nair *et al* (1999) noticed that neem formulations like neem azal (T/S 1% or 5%) 6 ml/litre, neem azal+ wettable sulphur 2 ml+3g/litre recorded 79.70, 75.10 reduction in infestation respectively. (Muthiah and Baskaran, 2000) reported that neem oil 2% + garlic extract 2% have effected 63 per cent population reduction.

Shivarama Reddy and Naik, (2000) has observed spraying of bunches at the crown with a mixture of neem oil 20ml mixed with garlic extract 20g and soap 50g at monthly interval effectively controlled the mite as observed in our study. Vidhyasagar (2000) stated that field trial carried out by Kerala Agricultural University,Thirussur showed that spraying either dicofol 6ml/litre two per cent neem oil +garlic mixture at monthly intervals, gave a reasonable control of the mite.

Muthiah and Baskaran (1999) found that spraying of neem oil 20 ml/litre at 10and 30 days interval effectively reduced the nut damage by the mites 30.20 and 43.50 per cent 90 days after treatment. (Subaharan et al, 2001) who observed that spraying of neem azal (T/S1% or 5%) 6 ml/litre of water recorded 79.68 reduction in infestation, spraying neem azal+ wettable sulphur 2 ml +3g/litre of water recorded 75.06 per cent reduction in infestation and also spraying of neem oil 20 ml +garlic extract 20g +soap 50g /litre of water recorded 68.47 reduction in infestation. Subaharan et al, (2001) who observed that spraying of neem azal (T/S1% or 5%) 6 ml/litre of water recorded 79.68 reduction in infestation, spraying neem azal+ wettable sulphur 2 ml +3g/litre of water recorded 75.06 per cent reduction in infestation and also spraying of neem oil 20 ml +garlic extract 20g +soap 50g /litre of water recorded 68.47 reduction in infestation.

1.4.3 Root feeding

Muthiah and Bhaskaran (1999) reported that root feeding of monocrotophos (10ml+10ml water) recorded 32.80 per cent of nuts damaged by mites after 90 days after treatment. Nair et al. (1990) reported that root feeding of monocrotophos (10ml+10ml water) recorded 81.10 reductions in infestation.

Root feeding experiments conducted at Avalur, Coimbatore district revealed that among the root feeding treatments monocrotophos (10ml+10ml water) was superior to all other chemicals by recording a maximum mortality of 61.57 and 73.55 per cent. Ramaraju et al (1999) also suggested that root feeding of monocrotophos 15ml once in 45 days satisfactory control is achieved. Shivarama Reddy and Naik (2000) reported that root feeding of 5-10 ml of monocrotophos twice mixed with equal quantity of water at 45 days interval was effective.

Vidyasagar (2000) reported that preliminary trials conducted by Central Plantation Crops Research Institute, Kasargode, Kerala have shown that administration of root feeding with monocrotophos (10ml+10ml water) at monthly intervals, adequate control of mite infestation.

Kannaiyan et al (2000) reported that two field experiments were conducted to evaluate the efficacy of Fish Oil Rosin Soap (FORS) 4% and other neem based botanicals viz., neem azal T/S 1% and 5ml/litre as spot application neem oil 3% and neem seed kernel extract 25% in 100ml as root feeding along with chemical insecticides. Sreerama Kumar and Singh (2000) suggested that root feeding with monocrotophos for the effective management of the coconut mite.

Shrama *et al* (2001) reported that root feeding of monocrotophos (10ml +10ml water) recorded 81.05 per cent reduction in infestation. Dey and Somchudhury (2001) observed that root feeding of monocrotophos (10ml +10ml water) recorded 62.62 per cent reduction of mite population at 8 days after treatment.

1.5 Seasonal incidence of *A.guerreronis*

Coconut mite attack was more severe in relatively dry climates or during dry season of wet climates (Zulaga and Sanchez, 1971). However, in other localities there was no clear relationship between coconut mite populations and wet and dry weather or if such relationship exists, it was obscured by other factors (Mariau, 1977; Howard, *et al.*, 1990; Ramaraju *et al.*,2000).

Julia and Mariau (1979) reported that in young bunches, the percentage of mite attack was greater during wet than in dry periods. Howard et al (1990) found that the coconut mite was seen in tropical and subtropical. Haq (1999a) studied the correlation between temperature and rainfall with the population of mites and found that the population density of the mite was positive correlated with the temperature and negatively correlated with rainfall. The population was reported to be maximum during summer or dry periods, even though the pest was present through the year (Nair and Koshy, 2000); Shivarama Reddy and Naik; 2000; Vidyasagar,2000). Studies on the weather relationship with eriophyid mite in coconut done at the Tamil Nadu Agricultural Meteorology, Tamil Nadu Agriculture University, Coimbatore revealed that maximum temperature had negative correlation with three months after the emergence of spathe. Flowers formed during rainy season escaped the attack of eriophyid mite (Swamiappan *et al.*, 2001).

1.6 Nature of damage caused by *A.guerreronis*

The mite infest and develop on the meristematic tissues of the growing nuts under the perianth by desapping the soft tissues of the buttons (Kannaiyan *et al.*, 2000). In the initial stage, the damage occurs as triangular patches close to the perianth as the nut grows in size, the feeding injury leads to warting and longituditional splits on the outer surface of the developing nuts Ramaraju, (1999). Due to this, yellowish triangular streaks descend down the three faces of the nut from the perianth (Ranjith *et al.*, 2001).

Feeding of the mites in the meristematic zone of the nuts causes physical damage so that as the newly formed tissues expand, the surface become necrotic and suberized (Moore and Howard, 1996). Draining of sap from young buttons causes reduction in the size of nuts (Estrada, *et al.*, 1975; Mariau,1986; Medina *et al*, 1986; Sathiamma *et al.*,1998) and leads to reduction in copra yield by 25 per cent (Mariau, 1977) and even up to 40 per cent (Muthiah and Baskaran,2000).

Premature nut fall was also reported in some areas due to the attack of mites (Mariau, 1977; Medina *et al.*, 1986) and this was disputed by Mariau (1986). Seguni (2000) reported that losses due to premature nut fall were between 10-100 per cent in Tanzania. In many cases, when perianth was removed, a pinkish band was seen on the inner side of button. At maturity the husk of infested nut was very tight and shrunken causing difficulty in dehusking (Nadarajan *et al.*, 2000).

1.7   Grading of damage symptoms by *A.guerreronis*

Based on the visible surface damage, Moore and Alexander (1978) classified the dry nuts into 5 damage categories Similar to that of Julia and Mariau (1979).

S.no	Symptoms	Percentage
1	Nuts with no mite damage	0
2	Nuts with superficial damage	1-10
3	Nuts with significant mite damage but not much smaller	11-25
4	Nuts with significant mite damage, smaller and with some distortion	26-50
5	Nuts heavily attacked, very much reduced in size and often greatly distorted	51-100

Similarly, Moore *et al.* (1989) classified the green nuts (about four months after fertilization) into 5 categories.

S.No	Symptoms	Percentage
1	None	0
2	Low	110
3	Medium	11-25
4	Severe	26-50
5	Very severe	51-100

Varadarajan (2000) has developed a 5-grade score that can be assess the damage to both the green and dry nuts

S.no	Symptoms	Grade
1	Plain and fresh nut surface without any injury	0
2	Scarification on the nut surface in random and / or triangular patches	1
3	Contiguous or discontigous scarification on the one fourth of the nut surface down the perianth	3
4	Contiguous or discontigous scarification on half of the nut surface down the perianth	5
5	Contiguous or discontigous scarification , either on the three-fourths of the nut surface with or without fissures and /or gummosis or on less than three-fourth of the nut surface with fissures and /or gummosis, with or without deformation	7

Kannaiyan *et al.* (2000) observed and allotted the following grades according to the extend of damage caused by *A.guerreronis*. The above nuts were categorized based on the external damage symptoms produced by the mite into four grades.

S.No	Percentage	Grade
1	No damage	0
2	25% damage	1
3	26-50 damage	2
4	>50% damage	3

MATERIALS AND METHODS

The present investigation entitled 'Studies on the management of coconut eriophyid mite, *Aceria guerreronis*, Keifer (Acari: Eriophyidae)" were carried out in the Department of Entomology, Faculty of Agriculture, Annamalai Univrsity, and at farmers coconut plantations in Kadavasal, near Chidambaram during the year 2000-2002.The materials used and methods followed in various experiments.

2.1 Survey on the incidence of *A.guerreronis* in the coastal belts of Tamil Nadu

A comprehensive survey on the percent infestation of nuts were conducted in Cuddalore, Nagapattinam, Thanjavur district of the coastal belts of Tamil Nadu. Two villages were selected per district. Virdhachalam and Kurinjipadi village of Cuddalore district. Mathanam and Memathur of Nagapattinam and Peravurani and Pattukottai of Thanjavur district.

Five palms were selected per village. The last matured bunch was selected for the survey. Totally thirty trees were surveyed throughout the year (September 2000 to August 2001).The trees were selected for survey was tagged. Percentage of nut infestation was worked out as per the following formula.

Percentage nut infestation = number of infested nuts per bunch
Total number of nuts per bunch

The trees exhibited the symptoms were rated as mild (0-10%+),medium (10-40%++) and Severe (>40%+++) (Kannaiyan *et al*,2000).

2.2 Studies on the bunch preference by *A.guerreronis*

Population of mites was supposed to vary with the age of the nuts. In order to identify the most preferred stage of nuts, a field experiments was conducted at the farmer field at Kadavasal, having uniform management practices. Ten palms of 15 years age of Tall x Dwarf variety were selected at random from a block. The experiment was laid out in randomized block design. From each palm, the nuts were collected randomly from the nine bunches at the rate of one nut per bunch, taking the youngest bunch with fertilized nut lets as bunch one. The flower bunch before, this was numbered as bunch zero. From this both male and female flowers were collected to check the presence of mites.

Collected nut were brought to laboratory and population of mite were determined by slightly modifying the "cellotape embedding Technique" to assess the

11

population of mites for more accuracy (Girija *et al* 2001).). In this technique, the perianth was removed from the button mechanically. A transparent cello tape of one inch width was taken and 8 mm^2 area was marked on the cello tape by using permanent marker pen. Then the cello tape was embedded on the nut surface. Population of mites was counted immediately after removing the perianth without disturbing the colony by keeping the slide under a Stereo microscope at 10X magnification. Counting was done using the hand tally counter. The mites that got adhered in the cello tape were counted to arrive at the total population of mites in 8 mm^2 square area.

2.3 Population dynamics of *A.guerreronis*

To study the population dynamics of mites in button, ten palms of Tall x Dwarf variety with age of 15 years were selected. The experiment used was randomized Block design. Based on the earlier experiments, the sampling was restricted to third bunch alone at the rate of one nut per bunch was taken. Observation on mite population were recorded by the method described in chapter 3.2.

2.4 Management of coconut eriophyid mite *A.guerreronis* on coconut

2.4.1 Plant products selected and prepared

Phytopalm, a herbal product which was used in liquid as well as in dust form were supplied by Hi-Tech Coconut Corporation, Nagercoil. The remaining selective botanicals neem azal (1000ppm), (P.J. Margo Private Limited), fortune aza (3000ppm) (Fortune Biotech Limited), neem seed, neem oil, dicofol 18.5% EC, monocrotophos 36% WSC were obtained from the market. Nochi leaf extract (3%), Calotropis leaf extract (5%) and neem seed kernel extract (5%) were prepared in the laboratory.

2.4.2 Preparation of plant extract

Phytopalm, a herbal product from Hi tech coconut corporation, Nagercoil was used as well as in dust forms. Both liquid and dust forms contain extracts and grounded forms of the following Custard apple, *Annona squamosa*, Purple tephrosia, *Tephrosia purpurea*, Kharanja, *Pongamia glabra,* Crown plant, *Calatropis gigantea*, Neem, *Azadirachta indica*, Garlic, *Allivum sativum*, Indian privet, *Vitex negundo* and Camphor), from this 30ml and 50ml was mixed with one litre of water to get 3% and 5% respectively.

Neem seed kernel extract

One kilogram of need seed was brought from Chidambaram market. The outer rind or seed coat was mechanically removes .the kernels were shade dried and ground into paste. About 50 grams of paste was added to one litre of water. This was kept undisturbed overnight and was filtered next day morning then 10 ml of teepal was added to it and sprayed.

Neem oil

Neem oil was obtained from the fertilizer shop,about 30ml of neem oil was mixed with one litre of water to get three per cent concentration. Ten ml of teepal was added to it and sprayed.

Nochi and calotropis leaf extract

Three and five kilograms of fresh nochi and Calotropis leaves were collected and cut into small pieces and thoroughly ground into paste. About thirty and fifty grams of paste was added to one litre of water to one litre of water to get three and five litre of concentration respectively. This suspension was kept undisturbed over night. Then next day morning, suspension was filtered through muslin cloth. Teepol was added to it and sprayed.

2.5   Laboratory bioassay

Laboratory bioassay was conducted to compare the bioefficacy of Phytopalm 5%, Phytopalm 3%, Neem azal 1%, Fortune Aza 1.5%, Neem Seed Kernel Extract 5%, Neem oil 3%, Nochi leaf extract 3%, Calotrpis leaf extract 5% against *A.guerreronis*. For comparing the efficacy of selected botanicals, monocrotophos 0.04% and dicfol was used as treated check. The bioassay was done in poison food technique. One month old nuts were taken from   un infested palms and meristematic zone of tender coconut was cut into small pieces (1cm.sq),dipped in 2ml of respective botanicals and insecticides solutions separately, air dried for 10 minutes and placed on wet cotton swab. Observations on mortality were recorded at 6.12.18 and 24 hours after treatment and per cent mortality was worked out (Dey *et al.*, 2001)

2.6 Management of mite *A.guerreronis* on coconut

2.6.1 Field trial

The present investigation was carried out in two consecutive seasons during September - December 2000 and January- April 2001 at Kadavasal Village, Chidambaram, Tamil Nadu using the variety Tall X Dwarf with age of 15 years. The inter and intra row spacing of 7 X 7 m and 10x10m. The treatments consisted of Phytopalm 5%, Phytopalm 3%, phytopalm contain the extracts of following herbs Custard apple, *Annona squamosa*, Purple Tephrosia, *Tephrosia purpurea*, Kharanja, *Pongamia glabra,* Crown plant, *Calatropis gigantea*, Neem, *Azadirachta indica*, Garlic, *Allivum sativum*, Indian privet, *Vitex negundo* and Camphor), Neem azal 1%, Fortune Aza 1.5%, Neem Seed Kernel Extract 5%, Neem oil 3%, Nochi leaf extract 3%, Calotrpis leaf extract 5% and monocrotophos 0.04% were evaluated against *A.guerreronis*. Monocrotophos 0.04% was used as standard check in randomized block design with three replications each replication consisting of one palm. From each of the selected palms, third bunch was selected from the top for spraying, since they have highest population. Spray application of pesticides on crown can be limited to third bunch alone (Ranjith *et al*, 2001). A schedule of three sprays was given at an interval of 30 days. The first spray was given during button stage when the infestation was noticed, pesticides were sprayed over the nuts of selected bunches; care was taken to drench the target area. Spraying was done in the early morning using rocker sprayer. Observations were taken on 1, 4 and 8 days after spray. The nuts were collected from the treated bunches and population of mites were determined by "cello tape embedding technique" (Girija *et al*, 2001). In this technique, the perianth was removed from the button mechanically. A transparent cello tape of one inch width was taken and 8 mm^2 areas were marked on the cello tape by using permanent marker pen. Then the cello tape was embedded on the nut surface. Population of mites was counted immediately after removing the perianth without disturbing the colony by keeping the slide under a Stereo microscope at 10X magnification. Counting was done using the hand tally counter. The mites that got adhered in the cellotape were counted to arrive at the total population of mites in 8 mm^2 square area. For comparison of all the treatments, Duncan's Multiple Range Test was adopted (Gomez and Gomez, 1984).

2.6.2 Management of coconut mite *A.guerreronis* through root feeding

Root feeding trials were conducted farmers field at Chidambaram, the first field trial was carried out during September-December 2001and second from January-April 2002. In both trials the variety used was Tall x Dwarf of 15 years old. The inter and intra row spacing was 7x7m and 10x10m respectively. The experiment was laid out in a randomized block design with ten treatments with two palms in each treatment which was replicated thrice. The treatments consists of phytopalm (20ml/palm), phytopalm (10ml/palm), neem azal (10ml/palm), fortune aza (20ml/palm) neem seed kernel extract (10ml/palm), neem oil (15ml/palm), nochi leaf extract (15ml/palm) and calotropis leaf extract (10ml/palm). Phytopalm a herbal product from Hi-Tech coconut corporation, Nagercoil, Tamil Nadu. It contains the extracts of the following herbs, Lantana, *Lantana camera* Linn., Custard apple, *Annona squamosa* Linn. Purple tephrosia, *Tephrosia purpurea* Linn., Kharanja, *Pongamia glabra* Linn., Crown plant, *Calotorpis gigantea* Ait, Neem, *Azadirachta indica* A.Juss.,G arlic, *Allium sativum* Linn., Indian privet, *Vitex nugundo* Linn. and Camphor. Monocrotophos (15ml/palm) was used as treated check. The extracts were mixed with equal quantity of water. For root feeding a pit was dug three feet away from the trunk in search of fresh roots. A freshly developed brick red coloured feeding root of pencil thickness was selected (Dey *et al*, 2001). A slanting cut was given to the root for exposure of vessels. A thick poly bag half litre capacity was filled with the above extract solution and the cut root was inserted into the polybag and tied air tightly with a thread to enable the cut portion to absorb the liquid. The next day morning all the liquid will be absorbed by the plant. The coconut water and kernel will be free from any residue of pesticide after 45 days of application. From each of the selected palms, the third bunch was selected for the population assessment.

After root feeding, one nut was taken from each palm and observations on number of mites 8mm^2 area at three places were recorded under Stereo binocular microscope and mean population was assessed. Observations on mite incidence were taken at (7, 15 and 23 days) after root feeding.

2.7   Seasonal incidence of *A.guerreronis*

The study was made from November 2000 to December 2000 in coconut at Kadavasal village. Field samples were collected at fortnightly intervals. The experiment was laid out in randomized block design with three replications, each replication comprising of one palm which was maintained unsprayed. In such selected trees, third bunch from top was selected. From the selected bunch, one nut was taken at random and observations on number of mites per $8mm^2$ area at three places were recorded and mean population was assessed.

Simultaneously, the meteorological data such as maximum, minimum temperature relative humidity, wind velocity, rainfall and sunshine hours were collected from the meteorological centre located at Annamalai University. Simple correlation and multiple regressions between three parameters and mite population were worked out (Swamiappan *et al.*, 2001).

2.8 Morphometric studies of *A.guerreronis*

Measurements of the egg, first, second instar nymphs and adults were done by using micrometry and the readings were recorded and tabulated (Ramerathinam and Loganathan, 2000).

Egg

The eggs were carefully removed from the nut by using a fine camel hair brush and kept in a glass slide and measured. The length is the distance from the tip of one end of the egg to tip of the other end. Width is the greatest distance measured from the outer margin f both sides. Measurement of 10 eggs was made and the arithmetic mean and standard deviation was calculated.

Nymphs

Ten nymphs were collected from the nut, then the first and second instar (Protonymph and deutonymph)were measured under microscope.

Adult

Adults were used for the measurement distance measured from the anterior most part of the head to the tip of the abdomen was considered as length and greatest distance across the body length as width respectively.

2.9 Nature of damage caused by *A.guerreronis*

Infested nuts of the palm were closely observed at various stages and nature of symptoms was elucidated.

2.10 Grading of damage symptoms caused by *A.guerreronis*

The visible surface damage that developed on the nuts of deferring maturity was taken. For this, green nuts were harvested from the palm. Based on the visible surface damage. The nuts were classified into 5 damage categories similar to the method as adopted by (Gomez and Gomez, 1984).

2.11 Statistical analysis

The data obtained from various experiments were suitably transformed into angular ($sin^{-1}/x \div 100$) values, if required and analysed using personal computer aided statistical package. For comparison of all the treatment means, Duncan's Multiple Range Test was adopted (Gomez and Gomez, 1984).Based on the observations recoded on seasonal incidence, simple correlation coefficients between mite population and weather parameters were worked out. Similarly, forecasting model was prepared by working out multiple regression equation, using computer aided statistical package. In morphological, bunch preference, population dynamics studies the data obtained were subjected to arthimatic mean and standard deviation.

EXPERIMENTAL RESULTS

The results of the investigations carried out during the study period at the Department of Entomology, Annamalai University and *A.guerreronis* at the farmers fields in Kadavasal, on the management of coconut eriophyid mite A*ceria guerreronis* Keifer (Acari: Eriophyidae) are presented below.

3.1 Survey on the incidence of mite in the coastal belts of Tamil Nadu

Severe damage was recorded at all the locations surveyed in Cuddalore, Nagapattinam and Tanjavur districts respectively during September 2000 to May 2001. But in Tanjavur district, this trend was continued up to August 200. In Cuddalore and Nagapattinam districts, the damage was reduced from June 2001 to August 2001, where the medium damage was noticed (Table.1)

Table.1 Incidence of *A.guerreronis* in the coastal belts of Tamil Nadu Sep2000 to 2001Aug)

Month	Cuddalore district		Nagapatttinam district		Thanjavur district	
	Vridhachalam	Kurinjipadi	Madhanam	Memathur	Peravurani	Pattukottai
Sep 2000	+++	+++	+++	+++	+++	+++
Oct 2000	+++	+++	+++	+++	+++	+++
Nov 2000	+++	+++	+++	+++	+++	+++
Dec 2000	+++	+++	+++	+++	+++	+++
Jan 2000	+++	+++	+++	+++	+++	+++
Feb 2001	+++	+++	+++	+++	+++	+++
Mar 2001	+++	+++	+++	+++	+++	+++
Apr 2001	+++	+++	+++	+++	+++	+++
May 2001	+++	+++	+++	+++	+++	+++
June 2001	++	++	++	++	++	++
July 2001	++	++	++	++	++	++
Aug 2001	++	++	++	++	++	++

+   Milld - (0-10%) ,++   Medium (10-40%), +++ Severe-(>40%)

3.2 Studies on the bunch preference by *A.guerreronis*

Mites can be seen on the interior side of the inner bracts corresponding to the mite colony inside the perianth where the population is high. Substantial number of mites was observed on the bracts covering area of attack inside the button. Mites were not found male and female flowers. Initially, white colored symptoms were found, second, third and fourth bunches, which indicates the initiation of infestation. The maximum population of 6.00mites/ 8mm^2 was observed in third, fourth and ninth bunches. It was noted that the mite population started declining from fourth bunch onwards (Table 2).

Table 2. Population of mite *A.guerreronis* in bunches of increasing maturity

Bunch Number	Population of mites/8mm2
	Mean±S.D
1	0±0
2	0±0
3	183.47±5.76
4	192.20±3.80
5	187.40±6.01
6	91.20±6.61
7	35.00±3.80
8	14.20±6.30
9	6.00±2.3

Mean values followed by standard deviation

3.3 Studies on population dynamics

Date on populations of mites was taken for a period of one year at monthly intervals from third bunch, which was fixed as index bunch. The results of the population dynamics were given in Table 3.

Table 3. Population dynamics of *A.guerreronis*

Month	*Population of mites/8mm2
	Mean±S.D
November 2000	188.29±2.93
December 2000	181.09±1.34
January 2001	187.16±2.47
February 2001	184.66±2.99
March 2001	191.73±3.12
April 2001	201.86±1.85
May 2001	217.15±5.56
June 2001	182.20±1.28
July 2001	180.03±1.92
August 2001	178.73±3.99
September 2001	172.86±1.93
October 2001	184.70±3.85

The results revealed that there were differences between the populations of mites at different intervals. The population of mites was high during April2001 (201.86 mites/8mm^2) followed by (217.15 mites/8mm^2) during May. The population started declining during June this trend continued up to September. There was a raise in population during October (184.70/8mm^2).

3.4 Bioassay I

Table 4.Efficacy of botanicals on mortality of *A.guerreronis* under laboratory conditions

S.no	Treatments	Per cent mortality hours after treatment)*			
		6h	12h	18h	24h
1	Phytopalm 1 g	13.33c	30.00c	40.00b	80.00bc
2	Phytopalm 1.5g	20.00bc	63.33a	70.00a	86.66b
3	Neem azal 1%	30.00ab	53.33ab	63.33a	76.66bc
4	Fortune Aza 1.5%	36.66a	46.66abc	63.33a	70.00cd
5	Neem seed kernel extract 5%	23.33bcd	43.33bc	60.00ab	66.66cd
6	Neem oil 3%	20.00bc	36.66bc	56.66ab	73.33cd
7	Nochi leaf extract 3%	30.00ab	33.33c	50.00b	52.33d
8	Calotropis leaf extract 5%	26.66ab	40.00bc	43.33b	60.00d
9	Monocrotophos 0.04%	40.00a	53.33ab	70.00a	100.00a
	Control	-	-	-	-
	S.D	4.16	5.25	5.72	4.87
	C.D (0.05%)	8.36	10.56	11.51	9.79

Mean values with different alphabets differ significantly

The results obtained from bioassay studies (Table 4) indicated that monocrotophos 0.04% ranked first after 24 hours of treatment with per cent mortality. This was followed by the botanicals like phytopalm 5% and phytopalm 3%both recording 86.66 and 80.00 per cent mortality respectively. Among the botanicals, Calotropis leaf extract 5% and nochi leaf extract 3% recorded low mortality rate of 60.00% and 52.33 per cent respectively.

3.4.1 Bioassay II

The results of the laboratory studies on the efficacy of various treatments against *A.guerreronis* (Table 5) showed that monocrotophos 0.04% was found significantly superior to all other treatments with 100 per cent mortality after 24 hours treatment. Neem oil 3%, phytopalm 1.5g, neem seed kernel extract 5% and nochi leaf extract 3% were on par with each other.

Table 5. Efficacy of botanicals on mortality of *A.guerreronis* under laboratory conditions

S.no	Treatments	Per cent mortality hours after treatment)*			
		6h	12h	18h	24h
1	Phytopalm 1g	16.66c	23.33c	40.00b	63.33c
2	Phytopalm 1.5g	20.00bc	33.33b	70.00a	80.00b
3	Neem azal 1%	16.66ab	30.00b	63.33a	56.66c
4	Fortune Aza 1.5%	33.33bc	40.00abc	63.33a	60.00c
5	Neem seed kernel extract 5%	20.00bc	33.33bc	60.00ab	63.33c
6	Neem oil 3%	23.33bc	40.00abc	56.66ab	70.00bc
7	Nochi leaf extract 3%	26.66abc	43.33ab	50.00b	56.66c
8	Calotropis leaf extract 5%	20.00bc	30.00bc	43.33b	63.33c
9	Monocrotophos 0.04%	40.00a	56.66a	70.00a	100.00a
	Control	-	-	-	-
	S.D	4.16	5.25	5.72	4.87
	C.D (0.05%)	8.36	10.56	11.51	9.79

Mean values with different alphabets differ significantly

The per cent mortality decreased in these above treatments to 70.00, 80.00, 63.33 and 56.66 per cent mortality after 24 hours. Neem formulations like neem azal 1% and fortune aza 1.5% were on par with each other. No mortality was observed on control after 24 hours spray.

3.5 Field efficacy of botanicals on per cent reduction of mite *A.guerreronis* (Field Trial I)

The data presented in the (Table 6) showed that on the 1st day after spray, Highest mean per cent reduction of mite population was noticed in monocrotophos 0.04%. During all the three spray phytopalm 5% and phytopalm 3%which were on par with each other.

21

Table 6. Per cent reduction of mite population *A.guerreronis* in various treatments (Trial I)

S.no	Treatments	* Percent reduction of mite population over control								
		Spray I			Spray II			Spray III		
		1DAS	4DAS	8DAS	1DAS	4DAS	8DAS	1DAS	4DAS	8DAS
1	Phytopalm 3%	10.38d	28.25c	32.58c	51.42c	55.52c	61.76c	52.94b	58.06c	53.93c
2	Phytopalm 5%	16.96b	36.59b	36.96b	59.04b	57.12b	75.48b	61.39a	60.59b	60.68b
3	Neem azal 1%	9.59cd	14.84e	24.10d	28.56e	42.67d	34.95g	22.78e	26.52f	36.06d
4	Fortune Aza 1.5%	9.96c	13.02e	24.58e	33.32d	36.59e	37.90f	25.37c	29.75e	32.10e
5	Neem seed kernel extract 5%	10.32c	18.47d	25.77e	23.81f	29.65g	41.17d	23.89d	26.52f	27.80g
6	Neem oil 3%	12.53b	18.83d	27.42d	24.76f	32.49f	39.21e	26.08c	30.00d	30.06f
7	Nochi leaf extract 3%	9.21d	13.08e	21.56f	14.58h	17.65i	30.25h	11.75g	18.99h	19.90i
8	Calotropis leaf extract 5%	11.43d	13.39e	22.38f	19.68g	24.28h	23.19i	15.06f	20.07g	21.03h
9	Monocrotophos 0.04%	32.85a	43.89a	49.94a	81.90a	84.85a	86.92a	61.77a	68.46a	73.81a
	Control	-	-	-	-	-	-	-	-	-
	S.D	0.34	0.53	0.89	0.48	0.63	0.25	0.29	0.27	0.27
	C.D (0.05%)	0.68	1.08	1.80	0.96	1.27	0.51	0.59	0.54	0.56

Mean values with different alphabets differ significantly, DAS-Days after Spray

Table 7. Mean per cent reduction of mite population *A.guerreronis* (Trial I)

S.no	Treatments	Mean per cent reduction of mite population		
		Days After Spray		
		1DAS	4DAS	8DAS
1	Phytopalm 5%	38.24b	51.43b	63.70b
2	Phytopalm 3%	45.92b	47.27c	62.60b
3	Neem azal 1%	20.31c	28.01d	32.77c
4	Fortune Aza 1.5%	22.88c	26.45d	32.70c
5	Neem seed kernel extract 5%	19.34c	24.88d	32.31c
6	Neem oil 3%	21.12c	27.10d	33.89c
7	Nochi leaf extract 3%	11.84d	16.57e	22.85d
8	Calotropis leaf extract 5%	15.39d	19.24e	20.99d
9	Monocrotophos 0.04%	58.71a	65.73a	77.92a
	Control	-	-	-
	S.D	3.68	3.24	1.67
	C.D (0.05%)	7.36	6.37	3.37

Mean values with different alphabets differ significantly, DAS-Days after Spray

The results presented in Table 7 showed that on fourth day after spray, monocrotophos 0.04% registered 65.73 per cent reduction followed by phytopalm 5% (51.43%) and phytopalm 3% (47.27%). Treatment with neem azal 1%, neem oil 3%, fortune aza1.5% and neem seed kernel extract 5% recorded 28.01, 27.10, 26.45 and 24.88 per cent respectively. On 8th day after spray, highest per cent reduction was observed in monocrotophos 0.04% (77.92%) next highest per cent reduction was registered in phytopalm 5% (63.70%) and phytopalm 3% (62.60%) followed by neem azal 1%, neem oil 3%, fortune aza1.5% and neem seed kernel extract 5% respectively.

3.6 Field efficacy of botanicals on per cent reduction of mite *A.guerreronis* (Field Trial II)

Mean per cent reduction A. guerreronis of first, second and third spray

The results presented in the Table 9 indicated that on the 1st day after spray maximum per cent reduction was observed with monocrotophos 0.04% (45.56%). Among the botanicals, phytopalm 5% and phytopalm 3% recoded 44.98% and 38.73% recorded in mite population. All the remaining treatments were significantly superior to untreated check. On the 4th day after spray, monocrotophos 0.04% recorded 60.76% per cent next higher per cent reduction was observed in the treatment *viz.,* neem oil 3%, neem seed kernel extract 5% fortune aza1.5% 25.02, 24.94 and 23.15 per cent respectively.

On 8th day after spray, it was found that among the botanicals tested phytopalm 5% (62.52%) and phytopalm 3% (57.73%) proved better than other treatments in reduction of mite population. The other botanicals in decreasing order of their efficacy were neem seed kernel extract 5% (33.50%), neem oil 3% (31.31%) and neem azal 1% (29.30).

Table 8. Per cent reduction of mite population of *A.guerreronis* (Trial II)

S.no	Treatment	* Percent reduction of mite population over control								
		Spray I			Spray II			Spray III		
		1DAS	4DAS	8DAS	1DAS	4DAS	8DAS	1DAS	4DAS	8DAS
1	Phytopalm 3%	16.60c	25.42c	49.88b	51.95c	58.73c	65.72c	50.69b	55.07c	57.55c
2	Phytopalm 5%	18.30b	35.47b	51.49b	58.49b	67.94b	75.87b	58.17a	59.42b	60.22b
3	Neem azal 1%	10.55c	17.19f	27.58d	28.84d	35.23d	34.20f	23.42c	24.64d	26.13d
4	Fortune Aza 1.5%	12.54d	15.76f	25.05e	25.48e	30.15d	31.47g	22.63de	23.55e	23.48e

S.no	Treatment	* Percent reduction of mite population over control								
		Spray I			Spray II			Spray III		
		1DAS	4DAS	8DAS	1DAS	4DAS	8DAS	1DAS	4DAS	8DAS
5	Neem seed kernel extract 5%	11.18d	20.47e	31.72c	19.27g	35.22d	47.22d	17.12f	19.20g	21.58f
6	Neem oil 3%	11.22d	22.65d	32.29c	21.22f	31.42e	37.78e	19.57d	21.01f	23.86e
7	Nochi leaf extract 3%	10.44d	12.54g	20.33f	13.72h	18.71f	22.78h	18.17d	18.11h	20.45g
8	Calotropis leaf extract 5%	10.51b	16.11f	18.26g	17.74g	20.00f	19.84i	16.07d	22.67f	14.70h
9	Monocrotophos 0.04%	20.66a	47.30a	70.00a	66.33a	72.38a	80.44a	58.70a	62.60a	65.90a
	Control	-	-	-	-	-	-	-	-	
	S.D	0.91	0.5	0.44	0.62	0.68	0.64	0.41	0.12	0.30
	C.D (0.05%)	1.83	1.32	1.27	1.24	1.37	1.30	0.83	0.25	0.61

Mean values with different alphabets differ significantly, DAS-Days after Spray

Table 9. Mean per cent reduction of mite population *A.guerreronis* (Trial II)

S.no	Treatments	Mean Percent reduction of mite population		
		1 DAS	4DAS	8DAS
1	Phytopalm 5%	44.98a	54.27b	62.52b
2	Phytopalm 3%	38.73a	46.50b	57.73b
3	Neem azal 1%	20.92b	25.68c	29.30d
4	Fortune Aza 1.5%	20.21b	23.15c	26.66d
5	Neem seed kernel extract 5%	15.85b	24.94c	33.50c
6	Neem oil 3%	17.33b	25.02c	31.31c
7	Nochi leaf extract 3%	14.12b	16.45c	21.18e
8	Calotropis leaf extract 5%	16.80b	19.59c	17.46e
9	Monocrotophos 0.04%	45.56a	60.76a	72.11a
	Control	-	-	-
	S.D	3.21	3.1	2.39
	C.D (0.05%)	9.31	6.42	4.84

Mean values with different alphabets differ significantly, DAS-Days after Spray

3.7 Root feeding of botanicals on per cent reduction of mite population of *A.guerreronis*

The results of the root feeding trial presented in the Table 10 showed that on the 7th day after root feeding, the percentage reduction was higher with monocrotophos @15ml/palm (17.80%) followed by phytopalm@ 20 ml/palm (10.61%),Calotropis leaf extract @10ml/palm (9.84%), neem azal @10ml/palm (8.71%), and fortune aza

@ 15 ml/palm (7.81%).Next higher per cent reduction was observed with neem oil @ 15ml/palm (6.43%), neem seed kernel extract@ 10ml/palm (6.06%) which were on par with each other.

On the 15th day after root feeding, monocrotophos @15 ml/palm (24.31%) gave higher per cent reduction followed by phytopalm @20 ml/palm (14.90%), neem azal @10ml/palm

Table 10. Mean per cent reduction in mite population *A.guerreronis* (First root feeding)

S.No	Treatments	Per cent reduction of mite population		
		7 DAR	15 DAR	23DAR
1	Phytopalm 10ml+10ml	4.54f	5.49g	8.33e
2	Phytopalm 20ml+20ml	10.61b	14.90b	18.33b
3	Neem azal 10ml+10ml	8.71c	12.55c	14.49c
4	Fortune Aza 15ml+15ml	7.81d	10.19d	8.33e
5	Neem seed kernel extract 10ml+10ml	6.06	6.66e	6.68f
6	Neem oil 15 ml+15ml	6.43e	9.02e	12.60d
7	Nochi leaf extract 15ml+15ml	3.02e	4.1g	8.33e
8	Calotropis leaf extract 10ml+10ml	9.84g	10.19b	10.83d
9	Monocrotophos 15ml+15ml	17.80a	24.31a	52.09a
	Control	-	-	-
	S.D	0.37	0.11	0.52
	C.D (0.05%)	0.74	0.22	1.06

Mean values with different alphabets differ significantly, DAS-Days after Spray

(12.55%), fortune aza @15 ml/palm (10.19%), Calotropis leaf extract @ 10ml/palm (10.19%), neem oil @ 15 ml/palm (9.02%) and neem seed kernel extract @ 10 ml/palm (6.66%) reduction. The minimum per cent reduction was recorded with nochi leaf extract@15 ml/palm (4.31%).On the 23rd day after first root feeding, maximum per cent reduction was observed with monocrotophos @ 15 ml/palm (52.09%). While treatments with phytopalm @20ml/palm,neem azal @10m l/palm and neem oil @ 15 ml/palm recorded 18.33,14.49 and 12.60 per cent reduction respectively. Phytopalm @15 ml/palm, fortune aza@15ml/palm and nochi leaf extract @15 ml/palm they were on par with each other.

Root feeding II

The per cent reduction of mite population after second root feeding are given in Table 11. On the 7th day after root feeding , monocrotophos @15ml/palm (28.89%) ranked first, this was followed by phytopalm @20 ml/palm (26.22%), neem azal @ 10ml/palm (17.46%) and fortune aza @ 15 ml/palm (10.22%).while treatment with phytopalm @15ml/palm recorded (6.66%), neem seed kernel extract @ 10ml/palm (7.52%) and Calotropis leaf extract @ 10ml/palm (7.08%). Minimum per cent reduction was observed with neem oil @15ml/palm and nochi leaf extract @15 ml/palm 4.55 and 2.20 per cent respectively.

Table 11. Percentage reduction in mite population *A.guerreronis* (Second root feeding)

S.No	Treatments	Per cent reduction of mite population		
		7 DAR	15 DAR	23DAR
1	Phytopalm 10ml+10ml	6.66e	13.34f	22.07d
2	Phytopalm 20ml+20ml	26.22b	36.53b	47.42b
3	Neem azal 10ml+10ml	17.46c	25.75c	33.34c
4	Fortune Aza 15ml+15ml	10.22d	17.35d	21.97d
5	Neem seed kernel extract 10ml+10ml	7.52e	15.06e	17.84e
6	Neem oil 15 ml+15ml	4.55f	6.39h	10.80g
7	Nochi leaf extract 15ml+15ml	2.20g	3.65i	9.85h
8	Calotropis leaf extract 10ml+10ml	7.08e	10.50g	15.02f
9	Monocrotophos 15ml+15ml	28.89a	44.75a	55.87a
	Control	-	-	-
	S.D	0.69	0.25	0.19
	C.D (0.05%)	1.40	0.51	0.38

Mean values with different alphabets differ significantly,

On the 15th day after root feeding, highest per cent reduction was observed with monocrotophos @15ml/palm (44.75%) followed by phytopalm @20ml/palm (36.53%), neem azal @10ml/palm (25.57%), fortune aza @15 ml/palm (17.35%),neem seed kernel extract @10ml/palm (15.06%), phytopalm @15 ml/palm (13.34%) and Calotropis leaf extract @10ml/palm (10.50%).lowest per cent reduction was observed with neem oil @ 15 ml/palm (6.39%) and nochi leaf extract @15 ml/palm (3.65%). On the 23rd day after root feeding, monocrotophos @15ml/palm recorded maximum reduction (55.87%) followed by phytopalm @20ml/palm (47.42%),neem azal @10ml/palm (33.34%).Treatments with phytopalm @15ml/palm and fortune aza @15ml/palm were on par with each other, among the botanicals, nochi leaf extract @15ml/palm was the least effective.

Root feeding III

Table 12. Percentage reduction in mite population *A.guerreronis* (Third root feeding)

S.No	Treatments	Per cent reduction of mite population		
		7 DAR	15DAR	23 DAR
1	Phytopalm 10ml+10ml	23.88d	24.11d	28.18de
2	Phytopalm 20ml+20ml	27.86b	30.26b	34.57b
3	Neem azal 10ml+10ml	21.84e	24.11f	26.58e
4	Fortune Aza 15ml+15ml	24.87c	25.68c	31.74bc
5	Neem seed kernel extract 10ml+10ml	17.42f	19.49g	28.18de
6	Neem oil 15 ml+15ml	21.89e	25.65c	30.13cd
7	Nochi leaf extract 15ml+15ml	6.97h	9.23g	14.37f
8	Calotropis leaf extract 10ml+10ml	10.95g	12.30f	21.70f
10	Monocrotophos 15ml+15ml	47.47a	53.34a	60.32a
	Control	-	-	-
	S.D	0.27	0.25	1.02
	C.D (0.05%)	0.54	0.51	2.06

Mean values with different alphabets differ significantly, DAS Days after Treatment

The results presented in the Table 12 showed that on the 7th day after root feeding, monocrotophos @15ml/palm recorded 47.47 per cent reduction in mite population followed by phytopalm @20 ml/palm (27.86%), fortune aza @15ml/palm (24.87%), phytopalm @15ml/plam (23.88%),neem oil @ 15ml/palm (21.89%), neem seed kernel extract @10 ml/palm (17.42%) and Calotropis leaf extract @10 ml/palm (6.97%).

On the 15th day after root feeding, monocrotophos @15ml/palm (53.34%) was found to be superior to all other treatments. Among the treatments, phytopalm @ 20ml/palm (53.34%) was found significantly superior to all other treatments. Among the botanicals, phytopalm @ 20 ml/palm recorded (30.26%) higher per cent reduction. All other treatments were significantly superior to untreated check.

On the 23rd day after treatment, monocrotophos @15ml/palm (60.32%) was superior to all in the reduction of mite population followed by

Table 13. Mean per cent reduction in mite population *A.guerreronis*

S.no	Treatments	Per cent reduction of mite population		
		7 DAR	15 DAR	23 DAR
1	Phytopalm 10ml+10ml	11.68c	14.28cd	19.60de
2	Phytopalm 20ml+20ml	22.45b	27.23b	33.64b
3	Neem azal 10ml+10ml	15.82bc	20.74bc	25.01c
4	Fortune Aza 15ml+15ml	14.30bc	17.73cd	20.79bc
5	Neem seed kernel extract 10ml+10ml	10.34c	13.71cd	15.23de
6	Neem oil 15 ml+15ml	10.95c	13.68cd	16.96cd
7	Nochi leaf extract 15ml+15ml	4.06d	5.73e	12.41f
8	Calotropis leaf extract 10ml+10ml	9.30c	10.99cd	19.51f
9	Monocrotophos 15ml+15ml	30.42a	40.80a	56.09a
	Control	-	-	-
	S.D	2.91	3.19	3.17
	C.D (0.05%)	5.86	6.41	6.38

Mean values with different alphabets differ significantly, DAS Days after Treatment

7 Days after root feeding

The results presents in the table 13 revealed that on the 7th day after root feeding, maximum per cent reduction was observed in monocrotophos @15ml/palm (30.42%). Next higher per cent reduction was observed in phytopalm @20ml/palm (22.45%).While treatments with neem azal @10ml palm and fortune aza @15 ml/palm recoded 15.82 and 14.30 per cent reduction respectively. They were on par with each other. All other treatments were more or less similar in their effects.

15 Days after root feeding

On the 15th day after root feeding, monocrotophos@ 15ml/palm (40.80%) performed well, followed by phytopalm @0ml/palm with 20.23 per cent reduction. Fortune aza @15 ml/palm, phytopalm @15ml/palm, neem seed kernel extract @ 10ml/palm and neem oil @15ml/palm recorded 17.73,14.28,13.71 and 13.68 per cent reduction respectively. They were on par with each other. Calotropis leaf extract@10ml/palm and nochi leaf extract @15ml/palm were less effective treatments.

23 Days after root feeding

On 23rd days after root feeding it was inferred that all the treatments were found to be superior over untreated check and inferior over treated check. Among the plant products, phytopalm @20 ml/palm showed highest per cent reduction (33.44%)

followed by all other treatments. Least per cent reduction of mite population was observed in nochi leaf extract @ 15 ml/palm (12.41%).

3.8 Second field trial

The results on efficacy of selected botanicals and an insecticide in each root feeding during second field trial are given below. Table 15

Root feeding I

On the 7th day after first root feeding, the percentage reduction was maximum with monocrotophos @ 15ml/palm (27.16%)and phytopalm 20ml/palm (17.39%)which were on par with each other followed by neem seed kernel extract @10 ml/palm (13.04%), Calotropis leaf extract @10ml/palm (12.68%). Neem oil @15ml/palm (11.95%) and neem azal @10ml/palm (10.86%) they were on par with each other (Table 14).

On the 15th day after root feeding, maximum per cent reduction was noticed with monocrotophos @15 ml/palm (31.11%) followed by phytopalm @20ml/palm (21.42%,) fortune aza @15 ml/palm (17.77%), neem azal @10ml/palm (16.66%),Calotropis leaf extract @10ml/palm (15.62%), neem seed kernel extract@10ml/palm (14.44%), phytopalm @15ml/palm (11.10%). nochi leaf extract @15ml/palm recorded the least per cent reduction (9.53%).

Table 15. Mean per cent reduction of mite population *A.guerreronis* (First root feeding)

S.No	Treatments	Per cent reduction of mite population		
		7 DAR	15 DAR	23 DAR
1	Phytopalm -10ml+10ml	11.59bcd	11.10h	11.11g
2	Phytopalm 20ml+20ml	17.39b	21.42b	18.67b
3	Neem azal 10ml+10ml	10.86b	16.66d	16.04d
4	Fortune aza 15ml+15ml	10.14cde	17.77c	17.28c
5	Neem seed kernel extract 10ml+10ml	13.04c	14.44f	13.58e
6	Neem oil 15 ml+15ml	11.95b	13.33g	14.81d
7	Nochi leaf extract 15ml+15ml	6.75e	9.53i	12.75f
8	Calotropis leaf extract 10ml+10ml	12.68c	15.62e	18.51b
10	Monocrotophos 15ml+15ml	27.16a	31.10a	47.24a
	Control	-	-	-
	S.D	0.96	0.38	0.22
	C.D (0.05%)	1.93	0.76	0.44

Mean values with different alphabets differ significantly, DAS Days after Treatment

On the 23rd day after first root feeding, monocrotophos was superior to all in reducing the mite population (47.24%), next phytopalm @20 ml/palm and Calotropis leaf extract @10ml/palm recorded 18.67 and 18.51 per cent reduction respectively. They were on par with each other followed by other treatments. Fortune aza @15 ml/palm, neem azal @10 ml/palm, neem oil @15 ml/palm, neem seed kernel extract @10 ml/palm, nochi leaf extract @ 15ml/palm and phytopalm @15 ml/palm recorded 17.28,16.04,14.81,13.58,12.75 and 11.11 per cent reduction of mite population respectively.

Root feeding II

On the 7th day after second root feeding, higher per cent reduction was observed with monocrotophos @15 ml/palm treated palms (29.83%). Next higher per cent reduction was observed in phytopalm 20ml/palm (24.33%),neem azal @10ml/palm(19.73%),fortune aza @15ml/palm (14.47%). Neem seed kernel extract @10 ml/palm, phytopalm @15 ml/palm and neem oil @15 ml/palm recorded 10.52, 10.52, 10.09, 9.21 and 7.89% reduction respectively.

Table 16. Per cent reduction in mite population *A.guerreronis* (Root feeding II)

S.No	Treatments	Per cent reduction of mite population		
		7 DAR	**15DAR**	**23 DAR**
1	Phytopalm 10ml+10ml	11.59bcd	11.10h	11.11g
2	Phytopalm 20ml+20ml	17.39a	21.42b	18.67b
3	Neem azal 10ml+10ml	10.86b	16.66d	16.04d
4	Fortune Aza 15ml+15ml	10.14cde	17.77c	17.28c
5	Neem seed kernel extract 10ml+10ml	13.04de	14.44f	13.58e
6	Neem oil 15 ml+15ml	11.95b	13.33g	14.81d
7	Nochi leaf extract 15ml+15ml	6.75e	9.53i	12.75f
8	Calotropis leaf extract 10ml+10ml	12.68b	15.62e	18.51b
10	Monocrotophos 15ml+15ml	27.16a	31.10a	47.24a
	Control	-	-	-
	S.D	0.96	0.38	0.22
	C.D (0.05%)	1.93	0.76	0.44

Mean values with different alphabets differ significantly, DAS Days after Treatment

On the 15th after second root feeding , monocrotophos @15ml/palm gave higher per cent reduction of 46.81 per cent followed by phytopalm @20 ml/palm (26.35%, neem azal @10ml/palm (24.99%), fortune aza @15 ml/palm (19.53%), nochi leaf extract @15 ml/palm (16.80% and phytopalm @15 ml/palm (15.59%). On the 23rd day after second root feeding, an increased per cent reduction was observed in treatments with monocrotophos aa@15 ml/palm (55.67%) and neem azal @10ml/palm (34.28%).while treatments with phytopalm @20ml/palm, phytopalm @15ml/palm, fortune aza@15 ml/palm , neem seed kernel extract @10ml/palm and nochi leaf extract @15 ml/palm recorded 28.57, 22.85,22.85,19.99 and 17.14 per cent respectively. calotropis leaf extract @10ml/palm recorded minimum per cent reduction (14.28%).

Root feeding III

On the 7th day after third root feeding, monocrotophos @15ml/palm recorded higher per cent reduction of (47.35%) followed by phytopalm @ 0ml/palm (31.88%),fortune aza @15 ml/palm (26.08%), neem azal @ 10ml/palm (22.07%), neem seed kernel extract @ 10ml/palm (18.85%),Calotropis leaf extract @ 10ml/palm (14.47%) and nochi leaf extract 10 ml/palm (11.59%). Among these neem oil 15 ml/palm and fortune aza @15 ml/palm were on par with each other. On the 15th day after third root feeding, monocrotophos@15 ml/palm (48.90%) stood first in reducing mite population followed by other treatments. Nochi leaf extract @15ml/palm was the least effective treatment. On the 23rd day after third root feeding , monocrotophos @15 ml/palm (51.98%)was found significantly superior to all other treatments followed by phytopalm @ 20ml/palm (35.35%), neem oil @ ml/palm (33.15%) reduction. All the treatments were more or less similar in their effect. Nochi leaf extract @15ml/palm (13.85%) recorded minimum reduction of mite population (Table 17).

Table 17. Mean per cent reduction of mite population after first root feeding (Field trial II)

S.no	Treatments	Per cent reduction of mite population		
		7 DAR	15 DAR	23DAR
1	Phytopalm 10ml+10ml	4.54f	5.49g	8.33e
2	Phytopalm 20ml+20ml	10.61b	14.90b	18.33b
3	Neem azal 10ml+10ml	8.71c	12.55c	14.49c
4	Fortune Aza 15ml+15ml	7.81d	10.19d	8.33e
5	Neem seed kernel extract 10ml+10ml	6.06	6.66e	6.68f
6	Neem oil 15 ml+15ml	6.43e	9.02e	12.60d
7	Nochi leaf extract 15ml+15ml	3.02e	4.1g	8.33e

S.no	Treatments	Per cent reduction of mite population		
		7 DAR	15 DAR	23DAR
8	Calotropis leaf extract 10ml+10ml	9.84g	10.19b	10.83d
9	Monocrotophos 15ml+15ml	17.80a	24.31a	52.09a
	Control	-	-	-
	S.D	0.37	0.11	0.52
	C.D (0.05%)	0.74	0.22	1.06

Mean values with different alphabets differ significantly, DAS Days after Treatment

7 DAR

On the 7th day after root feeding, it was noticed that monocrotophos @15 ml/palm (40.14%) gave the highest per cent reduction during all the three root feedings followed by phytopalm@ 20 ml/palm and neem azal @20 ml/palm recorded 24.53 and 17.76 per cent respectively. All the other treatments were equal in their efficacy, which were on par with each other. Nochi leaf extract@15ml/palm (10.75%) was the least effective treatment.

15 DAR

On the 15th day after root feeding, it was noted that monocrotophos@15ml/palm (43.30%) and phytopalm @20ml/palm (26.41%) mite population was reduced followed by other treatments. (Table 22)

23 DAR

On the 23rd day after root feeding, the maximum control of mite population was recorded in case of monocrotophos @15 ml/palm (44.93%).among the botanicals, phytopalm @20ml/palm (27.53%) performed well, nochi leaf extract @15 ml/palm recorded minimum reduction (17.74%)of mite population.

3.9   Seasonal incidence of *A.guerreronis* in relation to weather parameters

Observations were made on the incidence of *A.guerreronis* from the first fortnight of November 2000 to October 2001. The population of mite was seen throughout the study and reached its peak during the first and second fortnight of May 2001 (221.86 mites/mm^2), (217.56mites/mm^2) (Table 18) the occurrence was minimum during the first fortnight of October 2001 (76.66 mites/mm^2).

Correlation of mite population with weather parameters

The meteorological parameters such as maximum, minimum temperatures, relative humidity, rainfall, wind velocity, hours of sunshine were correlated with mite population and correlation coefficients obtained were revealed that maximum, minimum temperature and wind velocity were positively correlated. Relative humidity, rainfall and sunshine were negatively correlated. The increase in these parameters decreased the mite population. By using the regression equations presented in Table 19. Mean mite population /8mm^2 can be predicted for a given set of meteorological parameters. An unit increase in maximum temperature keeping other parameters constant resulted in increase in population of 1.31/8mm^2 area. In contrast, a unit increase in rainfall resulted in decrease in mite population of 0.83/8mm^2.

Table 19. Seasonal incidence of eriophyid mite *A.guereronis* during November 2000 to 2001

Month	Fort night	Temp. (°C)#		Relative humidity#	Wind velocity (Km/hr)#	Hours of bright sunshine#	Rainfall (mm)#	Mean mite population (mm)
		Max.	Mini.					
Nov	I	31.15	24.07	79.71	3.91	9.31	2.70	189.76
2000	II	29.62	23.54	82.40	4.55	8.57	17.01	186.86
Dec	I	29.20	22.94	82.00	5.10	7.60	2.07	188.23
2000	II	28.34	20.14	72.57	6.00	9.92	0.00	185.66
January	I	28.28	22.41	81.28	4.51	8.14	1.15	184.59
2001	II	28.20	21.71	79.00	5.18	7.31	0.00	187.83
February	I	29.98	21.48	50.17	3.28	8.27	0.00	191.83
2001	II	30.77	18.80	75.71	2.77	10.34	0.00	193.89
March	I	31.21	21.60	74.71	3.21	10.02	0.00	182.33
2001	II	32.14	21.67	74.85	2.95	9.64	0.00	190.39
April	I	33.14	25.08	76.85	4.84	8.86	0.00	201.79
2001	II	32.37	25.21	8.14	4.78	4.12	2.48	201.15
May	I	36.67	26.48	68.57	7.98	10.68	7.45	221.86
2001	II	37.27	27.67	64.42	8.91	0.30	0.08	217.56
June	I	32.41	25.41	75.00	5.11	2.68	9.35	209.46
2001	II	34.27	25.72	66.28	8.76	4.73	0.00	211.03
July	I	37.28	27.48	58.28	7.92	2.60	7.47	207.28
2001	II	36.85	26.42	60.71	.14	5.10	6.38	183.03
Aug	I	33.18	24.57	71.42	6.10	3.60	3.85	184.80
2001	II	34.98	26.00	66.71	6.80	4.20	9.07	181.73
Sep	I	36.41	25.42	64.85	5.87	5.10	9.05	178.62
2001	II	26.14	24.42	69.57	4.95	3.60	9.35	178.73
Oct 2001	I	32.34	24.87	76.42	3.87	3.58	4.67	176.66
	II	32.04	25.08	83.42	3.37	5.70	5.98	180.69

Values of fortnightly observations

Table 20. Simple correlation matrix between the incidence of *A.guerreronis* and weather factors

	X1	X2	X3	X4	X5	X6	X7
X1	1.000						
X2	0.733*	1.000					
X3	-0.387*	-0.137*	1.000				
X4	0.651**	0.692**	-0.294*	1.000			
X5	-0.108*	-0.199*	0.170*	-0.289	1.000		
X6	0.153*	0.432*	0.111*	0.163*	-0.245	1.000	
X7	0.492	0.410	-0.126*	0.489	-0.124	-0.110	1.000

X1-Maximum temperature (°C)

X2-Minimum temperature (°C)

X3-Relative humidity (%)

X4-Wind velocity- (km/hr)

X5-Rainfall (mm)

X6-Hours of bright sunshine

X7- mite population

Table 21. Prediction model for incidence of mite *A.guerreronis* during Nov 2000-Oct 2001

S. no	Infestation yard sticks	Regression Coefficient (b)						Intercept (a)	Regression equation	Coefficient of determination
		bx1	bx2	bx3	bx4	bx5	bx6			
1	Mite population /8mm^2	1.311	0.678	-0.182	1.701	-0.158	-0.825	116.1391	Y=116.1391+1.311X1 +0.678X2-0.182X3+ 1.701X4-0.158X5-0.825X6	0.361

X1-Maximum temperature (°C) Y=Mite incidence (8mm2)

X2-Minimum temperature (°C) R^2-Coefficient of determination

X3-Relative humidity (%)   b-Regression coefficient

X4-Wind velocity- (km/hr)

X5-Rainfall (mm)

X6-Hours of bright sunshine

X7- mite population

3.10 Morphometrics of eriophyid mite *A.guerreronis* on coconut

The morphometric parameters such as body length and width of an egg, nymphal instars, (protonymph, deutonymph) and adults on coconut were presented in Table 22.

The length and width of the eggs were 58.60μm and 41.25μm respectively. The length of the protonymph increased from 68.95 μm to 113.28 μm in dutonymph. Likewise the width also increased from 35.35 μm 43.05 μm. The length and width of the adults were 168.77 μm and 47.83 μm respectively.

DISCUSSION

Eriophyid mite *A.guerreronis* Keifer on coconut is becoming a serious problem in the recent times. Since 1998 onwards it has attained major pest status in three peninsular states of India. Kerela, Karnataka and Tamil Nadu (Sathiamma *et al.*, 1998: Mohanasundaram *et al.*, 1999; Narashima Rao., 2000). As coconut is one of the major plantation crops in Southern States of India. The mite pose a serious threat to the economy of the coconut growing countries. The estimated average loss of copra yield was found to be 10-15 per cent in Tamil Nadu (Moore and Howard, 1996).

Keeping these points in mind, the present investigation was carried out to study the distribution, bunch preference, population dynamics, nature of damage, seasonal incidence of eriophyid mites. Few selected botanicals, an acaricide and an insecticide were evaluated for their efficacy against this mite pest.

4.1 Survey on the incidence of eriophyid mite, *A.guerreronis* in the coastal belts of Tamil Nadu

From the results of the survey conducted during the present investigations, it was evident that the mite population was distributed along the coastal belts of Tamil Nadu such as Cuddalore, Nagapattinam and Thanjavur district as observed earlier also by Kannaiyan *et al.* (2000), Moore and Haward (1996) and Muthiah and Bhaskaran (1999).

In the present study, severe damage was noticed in Thanjavur district during September 2000-August 2000. The reason for such a severe damage in Tanjavur district may be due to the presence of older trees and weather factors. The reports of Muthiah and Bhaskaran (1999) revealed that severe damage was observed at Lakshathopu village in Pattukottai taluk of Thanjavur district which is in confirmatory with the present findings.

But in Cuddalore and Nagapattinam districts, severe damage was continued from September 2000-May 2001. The declining trend was observed after May in Cuddalore and Nagapattinam districts may be due to increase in rainfall.

4.2 Bunch preference by *A.guerreronis*

In the present study, it was noted that the mites were not found in male flowers of any inflorescence. Studies of Ranjith et al. (2000) also showed the absence of mites in male flowers. This may be due to the absence of soft tissues in male flowers where mites can feed. Besides the development cycle f mites extend more than 10 days (Mariau, 1977) and by that time the male flowers fall off. Hence it is unsuitable site for infestation.

There were no mites in unfertilized flowers. Moore and Alexander (1987) and Ranjith *et al.* (2000) also reported the absence of mites in unfertilized female flowers. The absence of mites from the unfertilized female flowers may reflect the tight adpression of bracts to the nut (Hall *et al.*, 1979). In female flowers, the meristematic tissues where the mites usually feed were well protected inside the perianth since the perianth was well protected inside the perianth. Since the perianth was tightly adpressed the gap between the perianth and the flower was very less and the mites cannot enter into the meristematic tissues.

As per the results of the present study, when the female flowers (buttons)open up for pollination, the attachment between the perianth and the buttons becomes less tight and loosened giving entry for the minute sized mite and made it easy to enter into the inter space between the perianth and button. This was in accordance with the results of Nadarajan, *et al* (2000) who observed that there was also disproportionate growth between the perianth parts and nut proper, so that entry of mite was facilitated. Moore and Alexander (987) reported that 20 per cent of first bunch found to contain mites. The present studies established the absence of mites in the first fully opened bunch consisting tender nuts. These findings are in conformity with the reports of Ranjith *et al* (2001).

In the present study, the infestation was observed from the second bunch onwards. The number of mites gradually increased and reached peak on third bunch and thereafter it decreases. A few numberof mites were noticed even on ninth bunch .Similar observations were made by Ranjith *et al.* (2000) who observed maximum population of mites on the third and fifth bunchs.

The present study revealed that the meristematic tissues of second, third and forth bunches were very soft, so that the mites can feed easily. Most probably the colonization starts either from the second or third bunches. The inflorescence opens at the rate of one per month. So at every month the bunch number changes i.e bunch two in one month becomes bunch three in next month.

The mites took approximately 10 days for their development (Mariau, 1977). They develop at an average of two to three number of mites that enter the buttons of second and third bunch may be only a few, but within month, after interrupted multiplication, they become enormous.

In the present study, higher number of mites was noticed in the third or fourth bunches depending on whether the initial colonization was on second or third bunches. This was n agreement with the findings of Ranjith *et al.* (2001).

As per the results of the present study, it was noted that when the button get older, the meristematic tissue gets hardened and it would be difficult for the mites for feeding and from fifth bunch onward a reduction in the number of mites was noticed. The mites were found to be less in second bunch when it was at the initial stage of development, whereas in eighth and ninth bunches, it was the remnants of previous stage mite population was observed. Similar observations were made by Ranjith *et al.* (2001).The present study revealed that $2^{nd}, 3^{rd}$, 4th and 5th bunches were more preferred by mites for colonization. But maximum number of mites were noticed in buttons of third bunches (Plate 8).

Hence the third, bunch was fixed as indexed bunch and further observations were restricted to these bunches only. This was similar to the results of Ranjith *et al.* (2001).

4.3   Population dynamics

The present study revealed that the mite population showed a difference in their occurrence during different months. However, higher population was noticed during the month of May 217.15 mites/8mm^2. Kanaiyan *et al* (2000) reported that the population of eriophyid mite was maximum even during rainy months October – December which was in confirmatory with the present study.

4.4   Efficacy of botanicals on mortality of *A.guerreronis* under laboratory condition

The results obtained from bioassay studies indicated that all the botanicals and pesticides tested against *A.guerreronis* revealed that monocrotophos 0.04% and dicofol 0.04% was highly and effective in killing the mites. This was in accordance with the reports of Shanabasavana *et al.* (1999) and Umamaheswari *et al.* (1999) as they reported that monocrotophos and dicofol was highly effective against spider mite,*Tetranychus neocaledonicus.*

Among the plant products tested phytopalm (3% and 5%) and phytopalm dust (1g and 1.5g) were more effective. The neem products like neem azal (1%), fortune aza (1.5%), neem seed kernel extract (5%), neem oil (3%) recoded higher per cent mortality.

38

4.5 Field efficacy of botanicals on per cent reduction of *A.guerreronis*

In the field trial taken up on coconut for the management of eriophyid mite, A.*guerreronis*, ten mites were choosen of which eight were botanicals, one synthetic chemical and a control, spraying was given when the infestation was noticed (one month after fertilization of flower) or button stage. Spraying was repeated two times at monthly intervals. This was in confirmatory with the reports of Hernandez (1977), Julia and Mariau (1979), Muthiah and Baskaran (2000), Mariau and Techibozo (1973).

The effectiveness of synthetic chemical and monocrotophos (0.03%) was reported by Mariau and Techibozo (1973), Shivarama Reddy and Naik (2000),monocrotophos 0.04% being superior to other treatments. It was also evident by the reports of Mariau and Techibozo (1973), Julia *et al.* (1979), Childers (1996), Fernando *et al.* (2000), Ramaraju *et al* (2000) who reported that the effectiveness of monocrotophos against eriophyid mite.

The reports of earlier workers suggested that effective control was achieved by the treatments with monocrotophos 0.04% at every three weeks. Hernandez (1977) reported that effective results were obtained with sprays with 2 ml monocrotophos at intervals of 15-20 days to the inflorescence and to nut less than three months old. This was in agreement with the present study.

It was also reported that monocrotophos was effective in reducing the population of *A. guerreronis* on coconut by Nair *et al* (1999), Muthiah and Baskaran (2000), Subaharan (2001) and Dey *et al.* (2001) reported that when the pesticides were applied directly on the crown region, after 8 days of treatment significantly highest per cent reduction of (70.51%) was observed in case of monocrotophos 250 ml/100 litre of palms, which is in accordance with the present findings.

The unimpeded damage and spread of the mite resulted in the use of variety of chemical acaricides and insecticides. The results from experiments in many countries indicated that chemical control is possible but many number of treatments and the quantities of chemicals required normally make it uneconomical but also dangerous to the environment and various life forms, including human and domestic animals. Chemicals are also inimical to most of the natural enemies of the pest. Resistance to the chemicals and resurgence of pest were also reported.

The advantage of using monocrotophos when compared to dicofol it is having less LD 50 value of 21 mg/kg whereas in monocrotophos it is 430mg/g. Dicofol belonging to organochlorine, non – synthetic pesticide, is very closely related to

DTT, a banned chemical. Due to these reasons it was not used and recommended on a large scale in the coconut ecosystem and also causes adverse effect on the environment and non-target organisms (Sree ramakumar and Singh, 2000)

In these circumstances lab and field experiments were conducted with phytopalm, a herbal product of Hi-tech coconut corporation along with some botanicals to identify the promising and effective results.

Among the botanicals tested phytopalm 5% and phytopalm 3% were more effective. This was confirmed by the reports of Ranjan babu (2001) who reported the efficacy of phytopalm against the yellow mite *Polyphagotarsonemus latus* on chillies and red spider mite, *Tetranychus cinnabarianus* on brinjal at 14 days after spray, phytopalm 2lit/ha were recorded 50.49 and 61.40 per cent reduction in mite population. This was in agreement with present study.

4.6 Efficacy of botanicals against *A.guerreronis* through root feeding

The results of the root feeding trials showed that monocrotophos (15ml+15ml of water) was found to be most effective. The treatment of monocrotophos through root feeding was reported by Ramaraju (1999). Mohanasundaram, *et al* (1999), Ramaraju *et al.* (2000)and Sreerama kumar and Singh (2000).

Ramraju, *et al* (1999) reported the effectiveness of mononcrotophos (15ml+15ml of water) was significantly superior to all other treatments by recording a maximum mortality of 61.57 and 73.55 per cent supported the present findings.

Although the reports of Nair et al. (1999), Muthiah and Baskaran (1999) and Subaharan *et al.* (2001), 32.80 and81.05 per cent revealed that root feeding of monocrotophos 10ml+10ml water recorded 81.10 reduction in infestation and also Dey *et al.* (2001) observed that root feeding of monocrotophos (10ml+10ml of water) recorded 62.62 per cent reduction of mite population 8 days after treatment which was in confirmatory with the present study. Of the selected botanicals tested, phytopalm (20ml+20ml of water) provided better control at 23 days after treatment. The reports of Kannaiyan *et al.* (2000) reported the effectiveness of neem azal, neeem oil, need seed kernel extract against eriophyid mite. All these reports were in concurrence with the present findings.

4.7 Seasonal incidence of eriophyid mite *A.guerreronis*

During the present investigation, it was noted that the mite population was started developing from month of November, reached a peak during May thereafter it started declining. Such a seasonal pattern of incidence was earlier reported by Haq (1999), Ramaraju (1999),Nair et al (1999), Nair and Koshy (2000), Kannaiyan, et al (2000) they reported that the mite population was maximum during March, April and May. Kannaiyan, et al (2000) and Swamiappan et al (2001) observed that fairly high mite population was found even during rainy months, September, October and November and this was in accordance with the present findings.

The reports of Zuluaga and Sanchez (1971) revealed that the coconut mite attack was more severe in relatively dry climate or during the dry season of wet climate. Haq (1999a) who stated that the population density of mites were positively correlated with temperature and negatively correlated with rainfall. Which is in accordance with the present findings however Mariau (1977), Howard et al (1990) and Ramaraju et al. (2000) reported that there was no clear relationship between wet and dry weather factors. This was contrary to the above findings.

4.8   Morphometric of eriophiyid *A.guerreronis*

The present description on the morphometrics of egg, two nymphal instars and adults revealed that the length and width of the deutonymphs were113.28µm and 43.05µ.Whereas it was reported to be 138.40 µmin length and 26.30 µm in width by Ramerathinam and Loganathan (2000). This is possibly due to the changes in nutritional and climate factors.

4.9   Nature of damage

The present study has revealed that both nymphs and adults were living inside the perianth as colonies by lacerating and sucking the sap from the meristematic tissue. Even though they feed mainly on the meristematic tissue. When the number was enormous, they can be seen on the perianth also.

The first symptoms to appear take the form of whitish, triangular patch with the base at the level of the tepals. At this stage, when the floral parts are lifted up, a white area was perceived with accumulation of thousands of mites at all the stages of their development. The triangular patch turns brown and the epidermis of the nuts become crackled. As the nuts grew, this injury on the nuts leads to warting to and longituditional fissures on the nut surface draining of sap from the nuts resulted in nut poor development of the nut, reduction in nut size. The above findings were in line with the observations of Mariau (1977), Sathiamma et al (1998), Haq (1999b), Ramaraju (1999), Kannaiyan et al.(2000) and Nadarajan et al. (2000).

4.10 Grading of damage symptoms caused by *A.guerreronis*

Julia and Mariau (1979), Moore *et al.*(1989), Varadharajan (2000), Kannaiyan *et al.*(2000) observed various degrees of external damage symptoms. According to the extend of damage, different grades were allotted. Similar results were obtained during the present investigation.

Summary

The results obtained from the above study are summarized below

1. Per cent incidence of mite population was more pronounced in Thanjavur districtthan Cuddalore and Nagapattinam district

2. There were no mites in male and female flowers and also on buttons of first bunch. Infestation was started from the second bunch onwards from the top population increased further and reached a peak in third bunch and then decreased. Hence, bunch three with maximum mite population was fixed as index bunch. So the third bunch was proved to be more preferred by mites.

3. Determination of mite population from buttons of third bunch from November 2000- October 2001 revealed that higher population was recorded during May 2001.

4. In the laboratory bioassay studies, monocrotophos 0.04%, dicofol 0.04%and among the botanicals tested, dust formulation of phytopalm 1g phytopalm 1.5g and phytopalm 3%, phytopalm5% were found to be the superior. Next to phytopalm neem azal performed well.

5. Monocrotophos 0.04% were found to be highly effective. Among the botanicals, phytopalm 5% and phytopalm 3% showed effective results. But their performance was only next to monocrotophos 0.04% in both field trials.

6. In the root feeding trials, monocrotophos (15ml+15ml of water) and botanicals

7. Seasonal incidence of this pest was found throughout the year.

8. Peak incidence was observed during April, May and June and started declining during July.

9. Prediction model has been developed by multiple regression analysis for estimating the mite incidence per $8mm^2$ area for a given set of weather parameters.

10. The nature of damage caused by *A.guerreronis* was whitish triangular patches, warting, longituditional splitting, necrotic brown patches, draining of resulting in reduction in nut size.

11. Various degrees of external damage symptoms with their corresponding grade points could be used to calculate the visible surface damage

ACKNOWLEDGEMENT

The authors highly thankful to the head Dr. Hendry Louis, HI-Tech coconut corporation, Nager kovil for providing fellowship and necessary facilities during the course of investigation

REFERENCES

Chiders, C.C.1996.Toxicological test methods for eriophyid mites. In: Linguist ,E.E., M.W. Sabilis and J. Bruin (eds) eriophyid mites-their biology, natural enemies and control, Elsevier Science Publi., Amsterdam, The Netherlands pp 422-199.

Chandrika Mohan and A. Josephrajkumar.2013.Understanding damage symptoms and management of coconut pests, Indian Coconut Journal, 10-15.

Daniel Sundarajan, D and Thulasidas, S. 1993 Botany of field crops. Macmillan India limited, New Delhi.pp275.

Dey, P.K. and Somchoudary, A.K. 2001. Bio-efficacy of fenpyroximate against eriophyid mite on coconut (*Aceria guerreronis* Keifer) and its impact on *Amblyseius spp* Pestology, 25 (10):6-10.

Dey, P.K., Sarkar, P.K., Gupta, S,K and Somchoudhry, A.K. 2001. Evaluation of fenazaquin against eriophyid mite, *Aceria guerreronis* Keifer on cocnut vis-à-vis its impact on *Amblyseius spp*. Pestology, 25.

Estrada, O.J., Gonzalez, A.M., Ortiz, J.E. and Avilla, M.G. 1975 Damage caused to coconut by *Aceria guerreronis* (Acarina: Eriophydae) in Cuba. Revista de Agricultura-Cuba, 8(2):30-34.

Fernando, L.C.P., Wickramananda, I.R and Artichigo,N. S. 2000 Studies on coconut mite, *Aceria guerreronis* In Sri Lanka. International Workshop on Coconut eriophyid mite, *Aceria guerreronis*, Coconut Research Institute, Lunwila, Sri Lanka.

Girija,V.K, Uthamasamy, K and Ambily Paul.2001. Cello tape embedding technique for assessment of population of coconut eriophyid mite, *Aceria (Eriophyis) guerreronis* Insect Environment 7(1):35

Gomez,A.K and Gomez,A.A.1984.Statistical procedure for Agricultural Research. John Wiley and sons, Singapore, p.680.

Hernandez, F. 1977 Distirbution of the coconut mite, *Aceria gurreronis* in peninsular India (*Eriophysis*) *guerreronis* on the coast of Guerrero. Agricultura Tecnia en Mexico, 4:23-38.

Howard, F.W., Rodriguez, E.A Denmar, H.A.1990. Geographical and seasonal distribution of the coconut mite, *Aceria guerreronis* (Acari: Eriophyidae), in Puerto Rica Florida USA. J.Agric. University Puetro Rica, 74:237-252.

Handa, S.K. 1999. Principles of pesticide chemistry. Agrobios (India) Jodhpur PP:56-160.

Julia, J. F. and Mariau, D. and Hall, R.1979. Nouvelles researches en cote d' Ivorie sur *Eriophyes guerreronis* K., acarian ravageur des noix du Oleagineux 34: 181-189.

Julia, J. F. and Mariau, D. and Hall, R.1979.New research in the Ivory Coast on *Eriophyses guerreroni*s K. a mite pest of coconut. Chamber of Commerce and Industry of Marseillus, p.13-16.

Keifer, H.H, 1985. Eriophyid studies. B-14. Calif. Dept. Agric., Bureau of Entomol., p20.

Kannaiyan S., Sabitha Duraisamy, Rabindra R.J, Ramakrishnan, G and Baskaran, P 2000.

Integrated packages for the management and control of coconut eriophyid mite P99-112. of migration and colonization of the coconut eriophyid mite *Eriophyes guerreronis* (Keifer) (Acari: Eriophyidae). Bulletin of Entomological Research, 77 (4):641-650.

Moore, D., Alexander, L. 1985.Coconut mite on Coconut. FAO Plant protection bulletin, 33(3):119.

Moore, DD., Alexander, L. and Hall, H.A.1989. The coconut mite *A. guerreronis* Keifer in St. Lucia: yield loss and attempts to control it with acaricide, polybutene and *Hirsutella* fungus. Tropical Pest management 35:83-89.

Moore, D. and Howard, F .W.1996.Coconuts.pp.561-570.In: Linquist, E.E. Sabilis, M.W. and J.(eds). Coconut eriophyid mites-their biology, natural enemies and control. Elseiver Science Publi., Amsterdam, the Netherlands.

Muthiah,C and Baskaran,R.1999.Screening of coconut genotypes and management of eriophyid mite, *Aceria guerreronis* (Eriophyidae: Acari) in Tamil Nadu. Indian Coconut Journal, 30:(6):10-11.

Muthiah,C and Baskaran,R.2000 Survey, bioecology and management of eriophyid mite on coconut. Paper presented in interactive Workshop on coconut eriophyid mite held on 19.05.2000 at the Tamil Nadu Agricultural University, Coimbatore, p99-112.

Nadarajan, L., Ranjith, A. M., Thomas, J., Bedevil, S. P. and Nair, G. M. 2000 Coconut perianth mite and its management. Tech. Bull., Kerela Agricultural University, Thrissur, Kerela, p. 10.

Narasimha Rao. B. 2000. Residue of triazophos in coconut water and kernel when administered through roots Pestology., 24 (1):2-4.

Nair,C.P., Rajan P., AbrahamV.A, Chandra Mohan,Murali Gopal, Mailvaganan. 1999. Studies on nut infesting eriophyid mite in coconut plantations. Central Plantation Crop Research Institute, Kasargod. Annual Report. p88-89.

Nair, C.P.R, Rajan.P and Namboothri, C.G.N.2011.Adoption of pest management strategies for sustainable production in coconut. Indian coconut journal.15-19.

*Nair, C.P.R. and Koshy, P.K. 2000. Studies on coconut eriophyid mites *Aceria guerreronis* Keifer in India Abs: International Workshop on Coconut mite (*Aceria guerreronis*), 6-8 January, Coconut Research Institute, Sri Lanka.p7.

Nair, M.K and Rajesh M.K.2001. Coconut production and productivity. Indian coconut journal. 32(2) p2.

Nampoorthi, K. U.K.1999. Coconut cultivation practices. Central Plantation Crops Researc Institute, Kasaragod, p.1.

*Olvera, F.S. and Foncosa, S.O. 1986. The mite causing "Coconut rust" in Veracruz, Mexico (Acarina: Eriophyidae) Folia Entomologica Mexicana 67:45-51.

Ramaraju.K.1999. Ecology of coconut mites. In Ecology based pest management. R.J Rabindra, S. Palanisamy, P. Karuppasamy, K. Ramaraju and R. Phillip Sridhar (eds.).Scroll, Coimbatore.pp158-163.

Ramaraju, K. Nadarajan, Sundra Babu. P.C and Murali Ragini, G.T. 1999. Management of coconut eriophyid mite, *Aceria guerreronis* in Tamil Nadu. J. Acarol., 14 (1&2):82-83.

Ramaraju, K., Nadarajan, K., Sundra Babu, P.C., Palanisamy, S and Rabindra, R.J.2000.Studies on coconut eriophyid mite *A. guerreronis* K. in Tamil Nadu, India. Paper presented in the International Workshop on coconut eriophyid mite.6-8, Jan. 2000, Sri Lanka, p7-8

Ramarethinam, S. and Marimuthu, S.1998. Role of IPM in the control of eriophyid mite, *Aceri guerreronis* (K). An emerging menace in the coconut palm in Southern India.Pestology,22(12):39-47.

Ramerathinam, S., Marimuthu, S and Manisegaran, N.V. 2000. Studies on th effect of *Hirsutella thompsonii* K. and a derivative in the control of coconut eriophyid mite, *Aceria (Eriophyses) guerreronis* (K) Pestology, 24 (2):3-8.

Ramarethinam, S, Marimuthu, S and Murugesan, N.V.2000b. An in vitro method for assessing the infectivity of *Hirsutella thompsonii* (F.) on coconut eriophyid mite, *Aceria (Eriophyses) guerreronis* (K),Pestology, 24(4):3-8.

Ramarethinam, S. and Lognathan,S. 2000. Biology of *Aceria (Eriophyses) guerreronis* (K) (Acari: Eriophyodea: Eriophyidae)-a perianth mite infesting coconut grooves in India. Pestology. 24 (1):6-9.

Ranjith, A.M., Vidya, C.V and Natarajan, L.2000 Distribution of population and bunch preferenc by *Aceria guerreronis* Keifer. Abs, international conference on plantation crops PLACROSYM XIV, 12-15 December, Hyderabad, pp.16.

Ranjith, A.M,Vidya, C.V. and Nadarajan, L. 2001. Population distribution of the perianth mite, *Aceria guerreronis* Keifer on coconut bunches, Insect Environment,7 (1)31-33

Rajan babu,K. 2001.Bionomics and management of yellow mite, *Polphagus tarsonemus latus* (Banks) (Acari: Tarsonemidae) on chillies, Capsicum annum L. and red spider mite, *Tetranychus cinnabarianus* (Bosid.) (Acari: Tertranychidae) on brinjal *Solanum melongena* L. M.Sc.(Ag.) Thesis, Annamalai University, Chidambaram

Sathiamma, B.1981. Mite fauna associated with coconut palm in Kerala. Pp. 11-14. In: Channabasuvana, G.P. (Ed.). Progress in Acarology in India. Acarological Society of India, Bangalore.

Sathiamma, B. Radhakrishnan Nair, C.P. and Koshy, P.K. 1998. Outbreak of nut infesting eriophyid -mite, *Eriophyses guerreronis* (K.) in coconut plantations in India. Indian coconut Journal 29 (2):1-3.

Schliesske, J.1998. On the gall mite fauna (Acarina: Eriophydae) of *Cocos nucifera* L. in Costa Rica. Nachrichtenblatt des Deustschen Planzenschulzienstes, 40 (8-9):124-127.

Seguni,Z. 2000. Incidence, distribution and economic importance of the coconut eriophyid mite, *Aceria guerreronis* Keifer in Tanzanian coconut based cropping systems. International workshop on Coconut Mites (*Aceria guerreronis*), 6-8 January, Coconut Research Institute, Sri Lanka p-10.

Shivarama Reddy, L., and Naik, S.L. 2000. Spread of eriophyid mites into the coconut gardens of Chitoor district, Andra Pradesh, Indian coconut journal, 30 (12):8-9.

Sreeramakumar,P. and Singh,S.P. 2000. *Hirsuitella thompsonii*: the best biological control option for the management of coconut mite in India. Indian Coconut Journal, 31: (5):11-17.

Subharan,K., Vidyasagar,P.S.P.V., Rohini Iyer and Sreeja.2001.Mite-a mighty barrier in coconut production Plant. Horti. Technology., 2(6):43-44.

Sawmiappan, M. Balasubramanian,T. N., Rabindra, R.J., Ramaraju, K. Selvaraju R., Geethalaksmi and Murali Arthanari .2001.Seasonal climate Vs Management of coconut eriophyid mite. Short courses on capturing the benefits of Climate forecast in Agricultural Management, 20-29 June.pp76-86.

Sharanabasava, H., Manjunatha,M. and Hachinal, S.G.1999. Interaction *Chrysoperla carnea* (Neuroptera:Chrysopidae) with botanicals and recommended pesticides used against spider mite, *Tetranychus neocaledonicus* (Acari: Tetrantchidae) Journal of Acarology., 14(1&2);10-15.

Singh,R.K. 2002. Coconut provides food, drink and raw materials. The Hindu Survey of Indian Agriculture, p109.

*Uaciquete, A., Rao,Y.P., Topper, C.P., Caligari, P.D.S., Kullaya, A.K, Shomari, S.H., Kasug, L.J., Masawe, P.A.L. and Mpunami,A.A.1998. Coconut pest and disease situation in Mozambique, Proc. of the International Cashew and coconut Conference: Trees for life-the key to development, 17-21, February 1997, Tanzania, pp.527-529.

Umamageswari,T., Sharmilabharathi, C, Kanagarajan, R., Arivudainambi, S. and Selavanarayanan, V.1999. Neem formulations and castor oil – A safe way to manage okra red spider mits. Journal of Acarology.,14 (1&2):77-79

*Varadarajan, M.K. 2000. Ecoclogy and management of coconut eriophyid mite, *Aceria guerreronis* Keifer. M.sc. Thesis, Tamil Nadu Agricultural University, Coimbatore, p.172.

Vidyasagar, P.S.P.V.2000 Eriophyid mites on coconut and their management. Indian Coconut Journal.,13 (2):15-16.

*Zuluaga, C.I and Sanchez, P.A.1971 Coconut scab, Colombia Acta Agronomica, 221 (3):133-139.

Biography

Dr. K. Balaji received his doctorate degree in Agricultural Entomology during 2007. He served as Senior Research Fellow at Tamil Nadu Agricultural University for three years. He has got research fellowship for MSc and Ph.D. programme. He has awarded Vallalar Endowment merit award for best performance in doctorate research during 2004 from Tamil Nadu government and Junior Scientist of the year award, 2010 from National Science Academy, New Delhi. His research interest on Management of Coconut eriophyid mite and Host plant resistance in sesame against shoot webber and capsule borer are well recognized and widely cited. He has obtained Post Graduate Diploma in Bioinformatics during 2007 from Bharathiyar University and he has also awarded Post Graduate Diploma in Agricultural Extension Management, MANAGE, Hyderabad. He has been life member of the various scientific bodies. He has written one book, he has participated and published 17 research articles in national and international journals, seminars/symposia. In June 2009, he joined the Department of Agriculture, Tamil Nadu as Agricultural Officer.

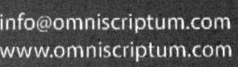

Printed by Books on Demand GmbH, Norderstedt / Germany